[零] A PHANTOM ZERO

RYU ANDO

the operating system c. 2019

the operating system digital print//document

[零] A PHANTOM ZERO

ISBN # 978-1-946031-66-2
copyright © 2019 by Ryu Ando
edited and designed by ELÆ [Lynne DeSilva-Johnson] with Orchid Tierney

is released under a Creative Commons CC-BY-NC-ND (Attribution, Non Commercial, No Derivatives) License: its reproduction is encouraged for those who otherwise could not afford its purchase in the case of academic, personal, and other creative usage from which no profit will accrue.

Complete rules and restrictions are available at:
http://creativecommons.org/licenses/by-nc-nd/3.0/

For additional questions regarding reproduction, quotation, or to request a pdf for review contact operator@theoperatingsystem.org

Print books from The Operating System are distributed to the trade by SPD/Small Press Distribution, with ePub and POD via Ingram, with production by Spencer Printing, in Honesdale, PA, in the USA. Digital books are available directly from the OS, direct from authors, via DIY pamplet printing, and/or POD.

This text was set in Steelworks Vintage, Europa-Light, Minion, Songti TC, Hiragino Mincho Pro, and OCR-A Standard.

Cover Art uses an image from "Collected Objects & the Dead Birds I Did Not Carry Home," by Heidi Reszies.

the operating system
www.theoperatingsystem.org
mailto: operator@theoperatingsystem.org

[零] A PHANTOM ZERO

A Phantom Zero

I. § N: The Drum Star (Orion's Ghost) —8
II. § R*: The R*umblings of Corrosion —18
III. § fp: The Specularium —28
IV. § ne: Potent Portents —33
V. § fl: Life in the lifeless things —38
VI. § fi: An Ode to Joy—43
VII. § fc: Broken Mirror / Sinking Ship—48
VIII. § L: Red Shifting To The Future—52

after-words—61

ACKNOWLEDGEMENTS

Section I of *A Phantom Zero* first appeared as "The Drum Star (Orion's Ghost),"
in *Strange Horizons*, September 2017, and in the *2018 Rhysling Anthology*.

Images adapted from *Tenmon Bun'ya No Zu: Chart of the Constellations and the Regions they Govern*. [天文分野之圖]National Astronomical Observatory of Japan.
Online at: https://www.nao.ac.jp/en/gallery/weekly/2014/20140313-old-illustration.html

Drake's Equation States:

$$N = R_* \cdot f_p \cdot n_e \cdot f_l \cdot f_i \cdot f_c \cdot L$$

"The definition of being is simply power."

-- Plato

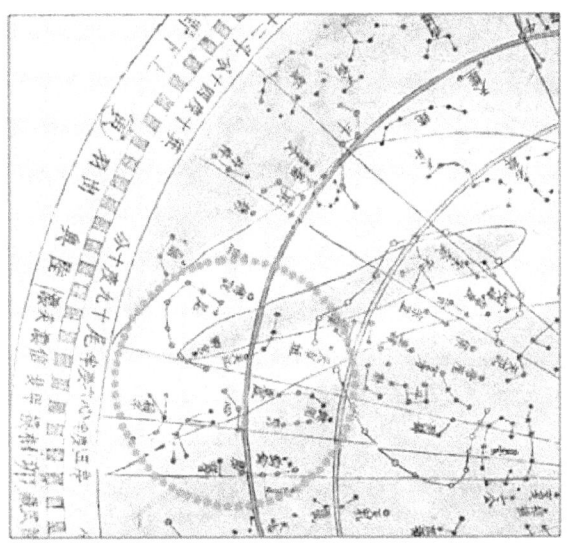

Detail from: T*enmon Bun'ya No Zu* 天文分野之圖

I. § N: The Drum Star (Orion's Ghost) *(Let N = the number of civilizations in our galaxy with which communication might be possible.)*

A. Finding the Tenmon Bun'ya No Zu: Chart of the Constellations and the Regions they Govern

Look inward to sky and heaven:

Our stories bind us
        To the stars,
    Bring us closer
        To their orbits,
            Their power and desire, their
    Consensual circles of
        Madness and fire.

Gravitas resides deep

    In those

Dark pockets

    Of the universe;

        Catharsis twists deeper

    Into the heart of our

        Own darkest

            Matters.

B. Poets of the infinite sea, drowned full fathom five

But there is no peace here,
>   Only ghosts and
>> Fragments of thought
>>> Scattered among your maps;
>> No hero lives in this poem.
>>> No hero lives
>>>> In this region of the sky.

>>> Look elsewhere for justice,
>>>> Look elsewhere for perfection,
>>> Not between these contested lines
>>>> And broken rhythms,
>>>>> Perpetually breaking in two.

As you write this

    Your ink dwindles, and

        The Universe expands,

            It covers you in flood;

        And you realize, too late, that

      Those are not pearls

                  That were his eyes.

B. Mathematical visions of mind-time

Fear is etched deep
With rune and quill,

sumi

And needle,
Into the arms
Of our slender galaxies,

– Violent ink writ large
Upon the menacing face of it all –
A spectral resonance haunts us.

Our suns, ancient wanderers,
    Have gone senile;
        Cold shoulders sag,
            Their lines strain toward
Blood-dimmed collapse
        Along the faults of our
            Wobbling dirt-filled empires.
Oedipal dreams and apocalyptic fevers
            Inhabit these skies;
We fill these rivers
        With blood-seeds and kill-spawn
        We cover these lands with
    Flowers of fire, inedible poisoned nectars,
        With cypress death and mad pomegranate,
            Writhing olive and black sakura.
Hungry poets, all, seize upon these demon seeds,
        Seize like sickly beasts,
            Speak of new lands un-blighted
    – Full of the same milk and honey,
Full of the same H_2O and DNA –
But dying in a tiny corner of the sky,
        Untouched.

D. The lost Teahouse of the Infinite Light

I remember you once waited
All night in the cold
      For Orion's ghost to rise,
Banging on the drum star
      tsuzumi boshi

And striking at the sky;
    But you did not see Him
        And in your hatred
    For the setting sun
        And the unsettling moon
Rising in silence and rage,
        In violence and pain,
You forgot to listen,
        Again.

There is no peace here,
    No universal hearth fire
        Comforting the chilled limbs
            Of your inner knowledge;
      Only horror   *and hate*   *and hiss*   *and loss*
The smell of the vain and your own glorious dead,
    The snap of your brittle flags planted
        On brittle maps with blurred legends,
    The plastic distractions
        Of the *Spiritus Mundi* gone mad,
Fueling you onward and outward,
    Strange beast,
        In an eternal *wonderlust*
            For permanence and order,
    For contact with mirrors.

E. The awakening light

Yet Orion's ghost shall rise again and
Bring the Universe to us,
Expanding and purging, as It does,
With light and flood – but only if you listen:

kenshō

Awaken

And we'll haunt these beautiful lines forever,
    – Suffer in a suspended sea-change –

        Shine on with the possibility of life,
    Dangle a future promise
        Bathed in the passing of light,
            Like the phantom stars
        And perfect circles that
            Haunt our skies.

II. § R*: The R*umblings of Corrosion

(Let R* = the average rate of star formation in our galaxy)

The great rumblings

    Of our fire-born mountains,

– Ideas born and then

    Flattened into driving,

Rhetorical plateaus,

    Only to rise again,

        In flame –

        Tell us that time

                Is but an offshoot

                            Of speed,

          A fiery tendril of shared experience

    Forged in an ache of death.

                  A fugue of life

Twists through us,

      A force of green-fused electric

                              Full of

        Intention wound forever tight

    Like a tourniquet.

Form is a mirage, the

    Mountains sigh through us.

       Their being is simply power:

色即是空

    shiki soku ze kuu

Suns are born

    In this quavering of time,

      And they die as

          Cold and numb as

    The years spent waning

      On that infinite plain.

All that will exist

Has already passed,

    A shadow crosses the dial of the sun;

        All remains

Suspended in stubborn

    Isolation,

    Locked in a quantum of amber:

        Atoms quarks bosons

   – Figural phantoms alike –

Fill the killing fields of time.

utakata no hibi

They shout as

A lotus perpetually

    Unfolds in lost

        Whispers;

A dissipating icicle sighs
    In eternal drip-drop
        Suspension;

A crow's arcing flight dives on
    Caught forever in
        Mid-air collisions;

A bubble spans
    In permanent ephemeral
        Tension, bursting

Dusk fades around
    A brightening star
        Sinking at our horizon

Ice burns my hands
    In a slow dissolution
        Of painful delight

Our sinking moon,

    Half lost in shadow,

        Hints at all the things

            We left unsaid

The writhing olive branch

    Held aloft in mid-breeze

        Sheds its half-moon leaves

And then a dream comes to me,

yume

Distant and fractured,

              Of sputtering pixel-red petals,

Decayed *sub rosa* projections

      Spilt upon wondrous

Hanging gardens of stone

Cracked black by fires

          Raging in eternal

     Extinguishment.

Seven billion wonders of the world

Collapsed upon themselves and

      Rose again in unison,

Since we last sat

           Together,

        Contracting and bulging in force.

We see the crushed remains

     Of our celestial calculations

Expanding and purging in time but

Lost to our time, an *Antikythera*

Groaning under the weight of

Corrosive centuries.

Salt scours our wounds, time

– *Unwound* –

Winds up yet again.

And yet a rose

Again blooms, sub rosa

Along these wasted, yet trembling

Solitary fringes

bara

And this impenetrable

    Fountain spills its secret:

        Take a piece of my petals,

        Folded like whispering lips,

    (It says)

            And give it, still quivering,

            So that you might taste

        The bitterness residing

        In all things beautiful.

            Incubate these tender growths,

            Explode them in rage and delight,

(It says)

Burn them in your cold embers,
Dim them in your long dotage,

> *Then speak to us in senile riddles*
> *About the infinite potential that's*

Always out of your reach; speak
In half-forgotten flower-spells

> *Suspended on the tips of your*
> *Ancient, secret, acidic tongues.*

III. § fp: The Specularium
(Let fp = the fraction of those stars that have planets)

Perhaps it is a lost

    And meaningless number

        That we seek,

            Like a phantom

                Zero:

rei

Brief windows of sky

    Look into future-time passing

        Through distant constellations.

But is this thing we seek whole?

                Or is it the patchy fragments and bones
Of something torn limb-from-limb,
        Piece-by-piece, like a horror-filled *Pentheus*?

Is it hiss and loss, black curse and spark,
        Drowned in the
                Wine-dark tinctures of
                      Radio silence,
                Collected in the deepest
        Pools of the universe?

Is this thing we seek
        Merely what lurks inside us,
                A pain masked by endless masks,
                          Leading to new pains?

            Is it merely conscious
                    Philosophies projected
                          Into dark caverns?

色即是空

shiki soku ze kuu

空即是色

kuu soku ze shiki

form, then, is emptiness,

and emptiness is form

When we speak of god, do

    We mean ourselves merely

        Thinking of thought itself?

    Do we enter

        These higher dimensions

            Thinking of thought itself?

Peer into the depths

    Of the infinite mirror as it

        Bends us backwards

            Through deepening time,

Into terror-filled temporal vistas

    (That too was godlike).

We see the threads of time

    (For time has bridled us together)

        Entwine us, weave a fabric,

            Unveiled as it binds us to ourselves

ito

And our stories trick us,

    Blind us to that lost number,

rei,

    Tell us we are

      Always on the

Verge of

    Discovery,

        Straddling the wire

At the revelatory edge

    Between truth and lie,

      Power and weakness, and

        The living and the dead.

IV. § ne: Potent Portents
(Let ne = the average number of planets that can potentially support life per star that has planets)

    Look beyond the

        Rising of the cypress

hinoki

Or the supple *ginko* or

被爆樹木

hibaku jumoku

                         Angled like rockets
            Toward the harsh, eternal blue
                     That spans, indifferent,
                              Above the clouds,
            Toward the bright lights
                         Lit up in the
                     Deep folds and recesses
                              Of the night.
                         Look beyond our possible pasts
            And impossible futures
                              Entwined as one with
                     The hope of Contact:

                                        Will we know It
                                        When we see it?

                                        Will It know us
                                        When It sees us?

I'm standing on the bridge

            Of heaven

                     Watching myself go by,

            Imprinted on its limestone bricks,

In an iridescent

Shinto-Tech atomic shadow,

    Clinging like crude matter

        To its perfect *eternal* forming.

A boiling river dissipates, *oleander*

    Full of blood and murder,

        Distills in a moment of time.

And we are unburdened

    By the atom

      – Its compassion

        Re-forms in our empty hearts –

Whose clock has now stopped,

    Zeroing in

On our frozen faces

        And twisted hands.

And we dream the dream of
The life we are living now

        It's been said, peering
                Into the other side
                        Of the window into time.

And when you want
To wake up, you will

        You'll pass into spirit:

rei

                Bathe in forests
Haunted by the likes of us.

Even if we forget the land,

        The land never forgets us.

For even

    In disintegration,

        Even in our darkest hours

The worm turns on us

    As we would turn upon it,

And we listen in rapture

    To the falling leaves,

        For those sounds as sullen,

            And intimate as the rain

V. § fl: Life in the lifeless things
(Let fl = the fraction of planets that could support life that actually develop life at some point)

    Heavy frost, full moon;
        Shocks on an
            Unexpected surface, and
    Light scatters.

            Perhaps something *was* here
    Swimming in these mud-
        Dimmed tides,
            But the oceans hold
                No memory of us.

    Like the portrait of a girl
        Embalmed in
            Daguerreotype and silver gelatin,
                Who later scratched out
                    Her own eyes:
        She sees the face
            Of the glass god
                Writ large across the sky,

    In the chill winds
        Of winters looming.

    And yet even there, the dust settles
Silent on the plain,
      Even beyond the horizon,
            On the solitary fringe
      Beneath the sky;
    Yes. Even *there* on
        That alien plain, beaten down
    Where no eyes can strain
        To apprehend this meaning
From meaning-lost men,
      The dust settles silent
          On the living and the dead.

            And these visions encompassed
              Over the alien plains
            Are thoughts gathered
                Like a handful of rain
              That drips into a mouthful of earth

These visions of babbling towers
Are now weeds gathered into a ball
Sent scattering over this vastness
Sinking into the pores of new lands.

I hear the sounds as intimate,
As sullen as the rain:

ame,

(O salve me)

And those opaque, glass gods,
Reveal their myth-laden faces
Among the

dekoboko

Rocks of dry riverbeds
　　　　　Holding
　　　　　　　　The fossil of all
　　　　Possible futures surely passed,
　　　　　　　　Those pieces of a
　　　　Cosmic puzzle,
　　　　　　　　　　Twisted.
But it *is* here,
　　　Can't you feel it?
　　　　　　Making the giant leap from
　　　　　　　　Lifelessness to life
Blowing through us in the
　　　　Twilight of our idols
　　　　　　　　And the low-lit spark
　　　　Of our *pareidolic* visions.

　　　　The joyless gods in their eternal
　　　　　　　　Madness rule us, just as we would,
　　　　　　　　　　　　Doubtless, have
　　　　　　　　　　　　　　Ruled ourselves.

Our chants breathe life
    Into their lifeless lives –
        I see their faces in the rocks,
            And in the uranium mines and
        In the distant shores
           Of distant worlds newly
                Imagined.

They tell me:

        Once seen,
            Nevermore un*seen.*

VI. § fi: An Ode to Joy

(Let fi = the fraction of planets with life that actually go on to develop intelligent life/civilizations)

       For one beautiful day,
            Through this flash
                Of indifferent blue,
       There was peace.

          So sing
                            <O Freunde>
       Sing us a new song.

           <Nicht diese tönen>

utau

Pray

    <O salve me>

        Bring us messages from
Beyond lifeless matter:

        Form is emptiness, emptiness form
            (Such words of wisdom).
        I am the pause between notes
            (The pause between thrusts)
                Which falls like the drip-drop of water.

        I am the conception of information
            And the inception of conformation;
                Two dimensions,
                    Rendered as flesh;
                        Three-dimensions in thought,
                Looping upward through
            An infinite stare.

                Emptiness becomes form.

I sit here

        (As eternity decays

                Into time and motion

                        And rises again)

                            Within

This massive expansion

                On a floor of *tatami*, bathed

                      In the infinite light,

       And think of

              How you, too, might

          Once have looked up at the stars at

                Night

            And thought of us.

Among the recursive heavens

          And unending loops of hell,

                We wonder what would have

Happened had

              Our shared lives lived on

    In perfect parallels.

       What would we be today?

            Would we have drunk in the stars

                As deeply as before?

Would you have turned us away
        Indifferent?

        Would we have
Destroyed you
        Even as we adored you?

        Time and tide lap at our feet
            Pull us inward,
                Closer to rhythm and rhyme,

<O sing
    Sing us a new song>

                As close as we can arrive
        Without sinking
             Into the purging flame
                Or the recursive wave or
                    The destructive mirror.

<O salve me>

Can you pray?

  (I hear the sounds
    Of sullen rain,
      Again)

      Let us pray:
  That the pixel images
    Of memory twisted like metal
  Towers scourged
    Will fade and scatter and

     Disperse
     Disperse

Like all the verses ever sung:

      O nicht diese tönen.

  (Let us pray)

    O salve me

VII. § fc: Broken Mirror / Sinking Ship
(Let fc= the fraction of civilizations that develop a technology that releases detectable signs of their existence into space)

    Dreams never end
            We barrel down
                Instead
Into their time-wells
        A multiverse hidden in our minds
            Looking up
      At the silver globe of
                Sky constrained above,
A sliver of night,
        Telescoped and tapered,
              Diminished and hushed, shimmering
                  Through the silence.

                Do we dare touch the sun?

All is writ in blood and thunder,
    Ailment and affliction;
        The seeker's curse and the poet's disease:
            We crawl across concrete
                      – Across hot coals and bent nails –
            On our hands and knees
        To arrive at the seeds of truth,
            Planted in alien soils, to
                Conjure the preconceived,
                    (And the god-eye peers inside).

    My own fingers, once soft,
        Are gnarled olive roots
            Twisting upon themselves.

And the damned souls
    Wandering in these olive groves
        Leap from my scrawl.

    The dip and quaff of the pen-scratch
        Releases the twisted
            From their torments,
                A resurrection in writhing
        Leafless thoughts.

The sun shines on me,
    A few minutes of light
  A day, an endless process of orbit
     And decay writ in *finite* terms.

And when it does,
    The mirror
  Of life shines into me;
Epiphany fills,
   Like water in the lungs; a
      Lost sunken ship
   Says it's time to reflect
     On the eternal possibilities
      Of the dead and the dying.

We long to ascend the clouds.
    Pyramids soaring in slopes
     To the stars, skyscrapers
   Gouging our blues, reforming the
  Ancient architectures
   Of babble and myth.

And the dreams never end.

                They just

    Mutate

        Like the sum of our

                Shared nightmares,

      Soaring outward

Into the abyss

              Of broken mirrors.

VIII. § L: Red Shifting To The Future
(Let L = the length of time for which such civilizations release detectable signals into space)

Rise of the radio phoenix

In a shift of red flame.

Radio waves collect

In the farthest corners

Of the universe.

What secrets lie hidden there?

Water echoes in a well,

The mystery wrapped within

Disintegrates

From three to two to one and the

Silence deafens in the eternal

Differences

Between the one and the many

The one and the zero.

kami-sama

                                       Is not merely a rule
            (Or a word
                  Or a thing
                          Or a logos)

It is the awakening
        Within that
                Moment of shift

Between our
        Strange loops
                When form is stripped
        Like the feathers from a paper crane
            Spanning the sky,
Wingless, bursting to flame, with
                The emptiness
                      Of a smokeless burn.

It is a transient lament
onozukara

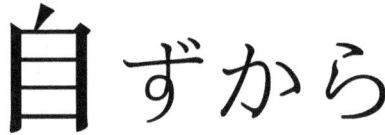

mizukara

As nature becomes self

And self, nature.

This is a beauty, then,
Which evolves through its own dissipation
A beauty which dissolves
In each breath I have taken
In each gasp I have stolen
In each time I have rested
For that finite moment of

Pause
Between my breaths.

As meaning is derived
            From new visions
    So the glass god is in *us*;
        Panoramic,
              Subatomic
    Sub rosa.

                  The fields of time
          Hold us all.

Hidden in power
    Hidden in strange currents
        Flowing strangely.

Escape the maze's loops,
    Derive life from the inert.

        Call forth the spiral in
    A fraction of time,
        Seconds are stretched out
            Like the pliant threads
                Of a tattered robe,

        Or an elegant tapestry
            Wrought in silken steel

                    Warped into
                Infinite looping dreams
        At just the moment
                    Before the chill arrives.

                        For we are swamped
                                By histories longer
                                        Than our own.

Will the hearth
            Signal us back home?

                    Will we ever
                            Sing a new song,

                                        <O freunde?>
                                Will we ever shift our thoughts,
        Escaping through layer
                    Upon layer through
        Trapdoors in the conscious mind
                    To find wisdom
                            In an infinite horizon
                                        Inhabited by our own?

      Will we revel

In the eternal word,

    And search for the great

          Reveal?

   There, ahead, embedded

       In mammoth rock

         We exist and haunt,

  Conjure ourselves as future specters

 And future shocks

      To haunt the coming ends.

   So don't ask 'why'.

      See what could be and

       Ask 'why not?'

          For the mind is the multiverse, and

      The multiverse, the mind.

And *there*,

   Through our recursive windows,

      There,

   Through the trapdoors of space-time,

There,

Through the *specularium*

Showing us the edge

Of this great expanse,

Like the milky tail

Of a dying comet,

Spanning the universe,

And then dispersing,

Disperse,

Disperse

There

Is peace,

(Perhaps,

One beautiful, fine day)

Within this verse.

心經

shingyou

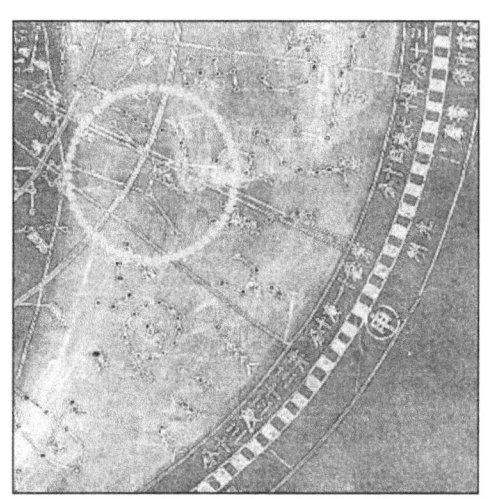

Tenmon Bun'ya No Zu 天文分野之圖

AFTER-WORDS

INFINITE / EMPTY : THE BRIGHT LIGHT OF ATTENTION
A CONVERSATION WITH RYU ANDO

Greetings! Thank you for talking to us about your process today!
Can you introduce yourself, in a way that you would choose?

I'm Ryu. I use this old-fashioned kanji: 龍
It means dragon.

Why are you a poet/writer/artist?

I'm a poet because I like writing poetry. It's fun for me. It also seems to convey what I want to say better than other genres or art forms. It's a bit of a mystery to me, though. I don't know where the desire or the poetry comes from. It spills out from someplace, often all at once.

When did you decide you were a poet/writer/artist (and/or: do you feel comfortable calling yourself a poet/writer/artist, what other titles or affiliations do you prefer/feel are more accurate)?

I had the sudden urge to "be a poet" when I was a teenager. So I wrote on/off for about 5 years, waiting for my moment of becoming something I wasn't. But then I stopped, unable to continue because it was totally "un-fun". After a break of about 20 years, I took up writing again, starting with short/flash fiction and wending my way back to poetry. That's when I realized I was ready, but to *do* not to be.

What's a "poet" (or "writer" or "artist") anyway?

I don't know, but I know one when I see one!

What do you see as your cultural and social role (in the literary / artistic / creative community and beyond)?

I want to bridge multiple ways of seeing the world, to mix them together, not only with languages and cultures, but also wider disciplines, such as the sciences and humanities. Science focused purely on 'advancement' for advancement's sake -- separated from ethics, philosophy, and art -- is dangerous. And art created purely for art's sake without looking at the astounding advancements of new knowledge is myopic.

Talk about the process or instinct to move these poems (or your work in general) as independent entities into a body of work. How and why did this happen? Have you had this intention for a while? What encouraged and/or confounded this (or a book, in general) coming together? Was it a struggle?

The pieces came together as a desire to look at the Drake Equation, which is a method for determining the possible extent of extraterrestrial life in the universe. It's a riff on each factor of the equation. I wrote the Drum star first from 2016 - 2017, and then worked on the remainder of these pieces for another year 2017-2018. I found old poems from early 1990s and extricated a few essential lines, essential themes and preoccupations from them that never went away. My process, though, remains unruly. I never know where it will go.

Did you envision this collection as a collection or understand your process as writing or making specifically around a theme while the poems themselves were being written / the work was being made? How or how not?

This was always intended to be a group of interlinked poems, each part altered and reshaped over time, reborn to work with the other pieces in the series while writing. What formal structures or other constrictive practices (if any) do you use in the creation of your work? Have certain teachers or instructive environments, or readings/writings/

work of other creative people informed the way you work/write?

A few things:
ma 間 [one of several concepts of space in Japanese aesthetics]: The concept of interval or gap, like silence in ambient music, shows up in the spaces between words, and line breaks. As Kenya Hara the designer suggests, "A space does not mean a volume defined by walls, but an area where consciousness brings the bright light of attention into play."

Ambient music / sound design: I'm obsessed lately with ambient music and how that can be used in poetry. Lately Yutaka Hirose's **NOVA**, but also Hiroshi Yoshimura's **Music for Nine Postcards**, vibraphonist Masayoshi Fujita, and Biosphere's **N-Plants**.

James Merrill states: "You hardly ever need to state your feelings. The point is to feel and keep the eyes open. Then what you feel is expressed, is mimed back at you by the scene." This speaks to my own process, of getting out in the world, to looking at my surroundings: recounting what I find in the sea and wave, sky and cloud, earth and tree, night and stars.

Speaking of monikers, what does your title represent? How was it generated?

零 **rei** is zero, nothing. It is also a pun on the word for spirit/ghost, which is also **rei**, but with a different kanji, 霊. Hence **A phantom zero;** but the circle, like the zero itself, is important in the concept of **ensō**, the moon in calligraphy, a symbol of the infinite, and a symbol of the emptiness.

Talk about the way you titled the book, and how your process of naming (individual pieces, sections, etc) influences you and/or colors your work specifically.

Each section is an elaboration, a thought experiment, on each section of the Drake Equation. Whether it actually hews closely to the formula itself is another story. I think of it as an improvisation on the equation. It's not very literal. It's a starting point.

What does this particular work represent to you as indicative of your method/creative practice? your history? your mission/intentions/hopes/plans?

Poetry is an ongoing process for me. I started working on the first few lines of this piece in 2016 and it expanded from that. I worked on sections and incorporated them as my 'gut' told me. It's all very intuitive and never planned out.

What does this book DO (as much as what it says or contains)?
It does nothing, like the inside of a circle.

What would be the best possible outcome for this book? What might it do in the world, and how will its presence as an object facilitate your creative role in your community and beyond? What are your hopes for this book, and for your practice?

My hope is that it inspires people to write their own works on whatever subjects speak to them.

Let's talk a little bit about the role of poetics and creative community in social and political activism, so present in our daily lives as we face the often sobering, sometimes dangerous realities of the Capitalocene. How does your process, practice, or work otherwise interface with these conditions?

All the great ideas in the world unleash forces beyond our control; the Drake Equation is revolutionary in that it takes seriously the idea that life **must** exist in other parts of the whole expanding universe -- which is itself a radical conception. To confirm it is to accept those forces of change, expansion, and the immense scale of time, which will wipe us all out. In that sense, the ideology that promotes the 'Market' as the solution to all things is very blinkered and but a blip in real-time. Talk about short-sighted! The word **hubris** also comes to mind. I hope that my work can diminish this **hubris**, at least a little, and help us see the world a bit more clearly, before we let it consume us.

I'd be curious to hear some of your thoughts on the challenges we face in speaking and publishing across lines of race, age, ability, class, privilege, social/cultural background, gender, sexuality (and other identifiers) within the community as well as creating and maintaining safe spaces, vs. the dangers of remaining and producing in isolated "silos" and/or disciplinary and/ or institutional bounds?

We must step outside ourselves and try to speak to everyone. It's a risk worth taking. Transcend boundaries. Unlock your door. Step outside. Speak what's true and people may listen.

ABOUT THE AUTHOR

RYU ANDO's writing has appeared in **Strange Horizons, Pidgeonholes, Liquid Imagination,** and other venues. His first book of poems, **The Lost Gardens of the Hakudo Maru**, is available from a...p press. Somewhere between L.A. and Saitama. This is where his characters exist and from where their voices carry. Lost and found. In Japan. In America. Sometimes both. Sometimes neither. Somewhere else entirely.

https://ryuando.wordpress.com

ABOUT THE COVER ART:

The Operating System 2019 chapbooks, in both digital and print, feature art from Heidi Reszies. The work is from a series entitled "Collected Objects & the Dead Birds I Did Not Carry Home," which are mixed media collages with encaustic on 8 x 8 wood panel, made in 2018.

Heidi writes: "This series explores objects/fragments of material culture—how objects occupy space, and my relationship to them or to their absence."

ABOUT THE ARTIST:

Heidi Reszies is a poet/transdisciplinary artist living in Richmond, Virginia. Her visual art is included in the National Museum of Women in the Arts CLARA Database of Women Artists. She teaches letterpress printing at the Virginia Commonwealth University School of the Arts, and is the creator/curator of Artifact Press. Her poetry collection titled *Illusory Borders* is forthcoming from The Operating System in 2019, and now available for pre-order. Her collection titled *Of Water & Other Soft Constructions* was selected by Samiya Bashir as the winner of the Anhinga Press 2018 Robert Dana Prize for Poetry (forthcoming in 2019).

Find her at heidireszies.com

WHY PRINT DOCUMENT?

The Operating System uses the language "print document" to differentiate from the book-object as part of our mission to distinguish the act of documentation-in-book-FORM from the act of publishing as a backwards-facing replication of the book's agentive *role* as it may have appeared the last several centuries of its history. Ultimately, I approach the book as TECHNOLOGY: one of a variety of printed documents (in this case, bound) that humans have invented and in turn used to archive and disseminate ideas, beliefs, stories, and other evidence of production.

Ownership and use of printing presses and access to (or restriction of printed materials) has long been a site of struggle, related in many ways to revolutionary activity and the fight for civil rights and free speech all over the world. While (in many countries) the contemporary quotidian landscape has indeed drastically shifted in its access to platforms for sharing information and in the widespread ability to "publish" digitally, even with extremely limited resources, the importance of publication on physical media has not diminished. In fact, this may be the most critical time in recent history for activist groups, artists, and others to insist upon learning, establishing, and encouraging personal and community documentation practices. Hear me out.

With The OS's print endeavors I wanted to open up a conversation about this: the ultimately radical, transgressive act of creating PRINT /DOCUMENTATION in the digital age. It's a question of the archive, and of history: who gets to tell the story, and what evidence of our life, our behaviors, our experiences are we leaving behind? We can know little to nothing about the future into which we're leaving an unprecedentedly digital document trail — but we can be assured that publications, government agencies, museums, schools, and other institutional powers that be will continue to leave BOTH a digital and print version of their production for the official record. Will we?

As a (rogue) anthropologist and long time academic, I can easily pull up many accounts about how lives, behaviors, experiences — how THE STORY of a time or place — was pieced together using the deep study of correspondence, notebooks, and other physical documents which are no longer the norm in many lives and practices. As we move our creative behaviors towards digital note taking, and even audio and video, what can we predict about future technology that is in any way assuring that our stories will be accurately told – or told at all? How will we leave these things for the record?

In these documents we say: WE WERE HERE, WE EXISTED, WE HAVE A DIFFERENT STORY

- Lynne DeSilva-Johnson [ELÆ], *Founder/Managing Editor,*
THE OPERATING SYSTEM, Brooklyn NY 2019

SELECTED RECENT AND FORTHCOMING OS PRINT/DOCUMENTS

ARK HIVE-Marthe Reed [2019]
A Bony Framework for the Tangible Universe-D. Allen [kin(d)*, 2019]
Y - Lori Anderson Moseman
Śnienie / Dreaming - Marta Zelwan/Krystyna Sakowicz,
(Polish-English/dual-language) trans. Victoria Miluch [glossarium, 2019]
Opera on TV-James Brunton [kin(d)*, 2019]
Alparegho: Pareil-À-Rien / Alparegho, Like Nothing Else - Hélène Sanguinetti
(French-English/dual-language), trans. Ann Cefola [glossarium, 2019]
Hall of Waters-Berry Grass [kin(d)*, 2019]
High Tide Of The Eyes - Bijan Elahi (Farsi-English/dual-language)
trans. Rebecca Ruth Gould and Kayvan Tahmasebian [glossarium, 2019]
I Made for You a New Machine and All it Does is Hope - Richard Lucyshyn [2019]
Illusory Borders-Heidi Reszies [2019]
Transitional Object-Adrian Silbernagel [kin(d)*, 2019]
A Year of Misreading the Wildcats [2019]
An Absence So Great and Spontaneous It Is Evidence of Light - Anne Gorrick [2018]
The Book of Everyday Instruction - Chloe Bass [2018]
Executive Orders Vol. II - a collaboration with the Organism for Poetic Research [2018]
One More Revolution - Andrea Mazzariello [2018]
The Suitcase Tree - Filip Marinovich [2018]
Chlorosis - Michael Flatt and Derrick Mund [2018]
Sussuros a Mi Padre - Erick Sáenz [2018]
Sharing Plastic - Blake Nemec [2018]
The Book of Sounds - Mehdi Navid (Farsi dual language, trans. Tina Rahimi) [2018]
In Corpore Sano : Creative Practice and the Challenged Body [Anthology, 2018];
Lynne DeSilva-Johnson and Jay Besemer, co-editors
Abandoners - Lesley Ann Wheeler [2018]
Jazzercise is a Language - Gabriel Ojeda-Sague [2018]
Return Trip / Viaje Al Regreso - Israel Dominguez;
(Spanish-English dual language) trans. Margaret Randall [2018]
Born Again - Ivy Johnson [2018]
Attendance - Rocío Carlos and Rachel McLeod Kaminer [2018]
Singing for Nothing - Wally Swist [2018]
The Ways of the Monster - Jay Besemer [2018]

THE 2019 OS CHAPBOOK SERIES

PRINT TITLES:

Vela. - Knar Gavin
[零] A Phantom Zero - Ryu Ando
Don't Be Scared - Magdalena Zurawski
Re:Verses - Kristina Darling & Chris Campanioni

✳✳✳

DIGITAL TITLES:

American Policy Player's Guide and Dream Book - Rachel Zolf
The George Oppen Memorial BBQ - Eric Benick
Flight Of The Mothman - Gyasi Hall
Mass Transitions - Sue Landers
The Grass Is Greener When The Sun Is Yellow - Sarah Rosenthal & Valerie Witte
From Being Things, To Equalities In All - Joe Milazzo
These Deals Won't Last Forever - Sasha Amari Hawkins
Ventriloquy - Bonnie Emerick
A Period Of Non-Enforcement - Lindsay Miles
Quantum Mechanics : Memoirs Of A Quark - Brad Baumgartner
Hara-Kiri On Monkey Bars - Anna Hoff

✳✳✳

PLEASE SEE OUR FULL CATALOG
FOR FULL LENGTH VOLUMES AND PREVIOUS CHAPBOOK SERIES:
HTTPS://SQUAREUP.COM/STORE/THE-OPERATING-SYSTEM/

THE 2019 SERIES MARKS OUR 7TH AND FINAL SPRING 4-BOOK SERIES
THANK YOU TO ALL THE WONDERFUL CREATORS BEHIND THESE TITLES

CHAPBOOK SERIES 2018 : TALES
Greater Grave - Jacq Greyja; Needles of Itching Feathers - Jared Schlickling;
Want-Catcher - Adra Raine; We, The Monstrous - Mark DuCharme

CHAPBOOK SERIES 2017 : INCANTATIONS
featuring original cover art by Barbara Byers
sp. - Susan Charkes; Radio Poems - Jeffrey Cyphers Wright;
Fixing a Witch/Hexing the Stitch - Jacklyn Janeksela;
cosmos a personal voyage by carl sagan ann druyan steven sotor and me - Connie Mae Oliver

CHAPBOOK SERIES 2016: OF SOUND MIND
*featuring the quilt drawings of Daphne Taylor
Improper Maps - Alex Crowley; While Listening - Alaina Ferris;
Chords - Peter Longofono; Any Seam or Needlework - Stanford Cheung

CHAPBOOK SERIES 2015: OF SYSTEMS OF
*featuring original cover art by Emma Steinkraus
Cyclorama - Davy Knittle; The Sensitive Boy Slumber Party Manifesto - Joseph
Cuillier; Neptune Court - Anton Yakovlev; Schema - Anurak Saelow

CHAPBOOK SERIES 2014: BY HAND
Pull, A Ballad - Maryam Parhizkar;
Can You See that Sound - Jeff Musillo
Executive Producer Chris Carter - Peter Milne Greiner;
Spooky Action at a Distance - Gregory Crosby;

CHAPBOOK SERIES 2013: WOODBLOCK
*featuring original prints from Kevin William Reed
Strange Coherence - Bill Considine; The Sword of Things - Tony Hoffman;
Talk About Man Proof - Lancelot Runge / John Kropa;
An Admission as a Warning Against the Value of Our Conclusions - Alexis Quinlan

DOC U MENT
/däkyəmənt/

First meant "instruction" or "evidence," whether written or not.

noun - a piece of written, printed, or electronic matter that provides information or evidence or that serves as an official record
verb - record (something) in written, photographic, or other form
synonyms - paper - deed - record - writing - act - instrument

[Middle English, precept, from Old French, from Latin documentum, example, proof, from docre, to teach; see dek- in Indo-European roots.]

Who is responsible for the manufacture of value?

Based on what supercilious ontology have we landed in a space where we vie against other creative people in vain pursuit of the fleeting credibilities of the scarcity economy, rather than freely collaborating and sharing openly with each other in ecstatic celebration of MAKING?

While we understand and acknowledge the economic pressures and fear-mongering that threatens to dominate and crush the creative impulse, we also believe that **now more than ever we have the tools to relinquish agency via cooperative means,** fueled by the fires of the Open Source Movement.

Looking out across the invisible vistas of that rhizomatic parallel country we can begin to see our community beyond constraints, in the place where intention meets
resilient, proactive, collaborative organization.

Here is a document born of that belief, sown purely of imagination and will.
When we document we assert. We print to make real, to reify our being there.
When we do so with mindful intention to address our process, to open our work to others, to create beauty in words in space, to respect and acknowledge the strength of the page we now hold physical, a thing in our hand… we remind ourselves that, like Dorothy: *we had the power all along, my dears.*

THE PRINT! DOCUMENT SERIES

is a project of
the trouble with bartleby

in collaboration with
the operating system

www.ingramcontent.com/pod-product-compliance
Lightning Source LLC
Chambersburg PA
CBHW080636130526
44591CB00047B/2704